Bloodsuckers

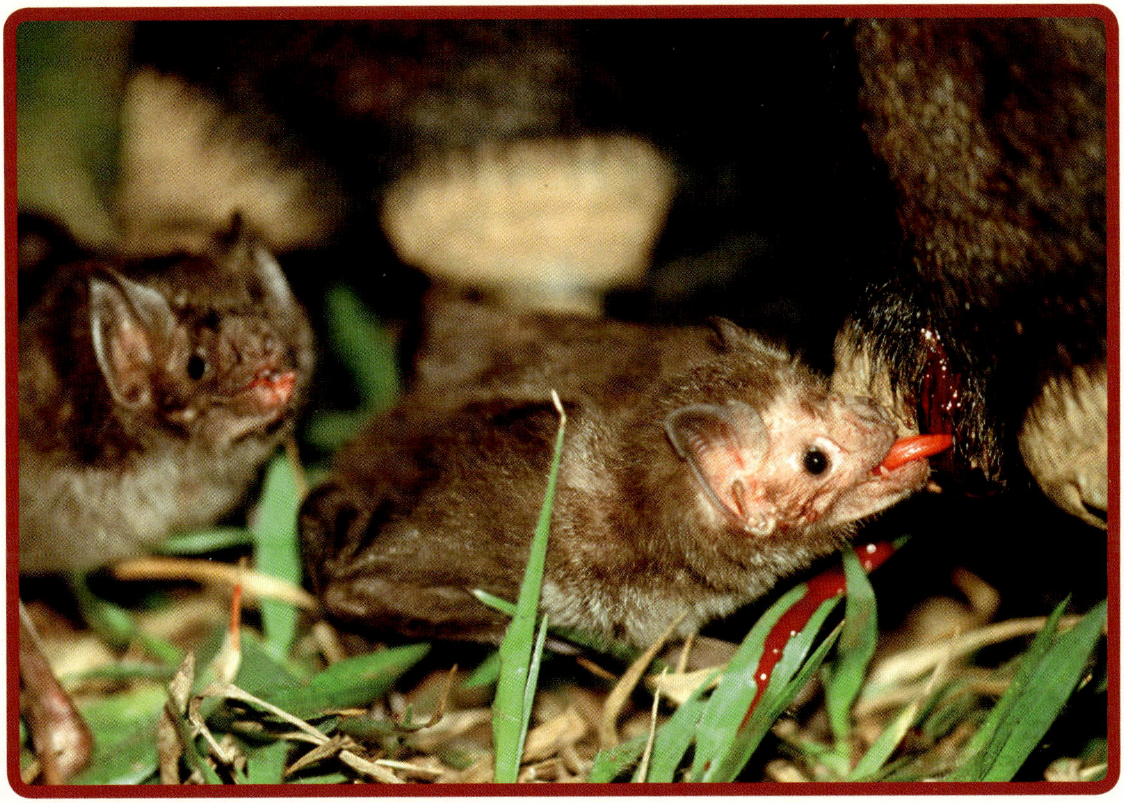

David and Patricia Armentrout

Vero Beach, Florida 32964

© 2009 Rourke Publishing LLC

All rights reserved. No part of this book may be reproduced or used in any form or by any means, electronic or mechanical, including photocopying, recording, or by any information storage and retrieval system without permission in writing from the publisher.

www.rourkepublishing.com

PHOTO CREDITS: © Michael Pettigrew: Cover Image; © Photolibrary/ Nick Gordon: Title page, page 12; © Nathan Jones: page 4; © Andrei Nekrassov: page 5, 16; © Chris Schuster, © Lawrence Sawyer: page 6; © Mircea Bezergheanu: page 7, 8; © Troy Casswell: page 9; © U.S. Fish and Wildlife Service/ Andy King: page 10 © Michael Lynch: page 11; © Craig Smith: page 15; © Herb Sennet: page 15; © Dave Brenner/ Michigan Sea Grant: page 17; © U.S. Fish and Wildlife Service: page 18; © EPA: page 19; © James Gathany: page 20, 21, 23, 24, 25; © Piotr Naskrecki: page 26; © Dr. Harold Harlan: page 27; © Kurt_G: page 28; © Dr. Morley Read: page 29

Editor: Kelli Hicks

Cover Design: Nicola Stratford: bdpublishing.com

Page Design: Renee Brady

Library of Congress Cataloging-in-Publication Data

Armentrout, David, 1962-
  Bloodsuckers / David and Patricia Armentrout.
     p. cm. -- (Weird and wonderful animals)
  ISBN 978-1-60472-301-4 (hardcover)
  ISBN 978-1-60694-907-8 (softcover)
  1. Bloodsucking animals--Juvenile literature. I. Armentrout, Patricia, 1960-
II. Title.
  QL756.55.A76 2009
  591.5'3--dc22
                                          2008019693

Rourke Publishing
Printed in the United States of America, North Mankato, Minnesota
120110
113010LP-A

www.rourkepublishing.com - rourke@rourkepublishing.com
Post Office Box 643328 Vero Beach, Florida 32964

Table of Contents

Vampires	4
Bloodsucking Worms	6
Vampire Bats	10
Calling Dr. Maggot	14
Sucker for a Pretty Face	16
Ticks	20
Deadly Bloodsuckers	22
Don't let the Bedbugs Bite!	26
Killer Bugs	28
Tools of Nature	30
Glossary	31
Index	32

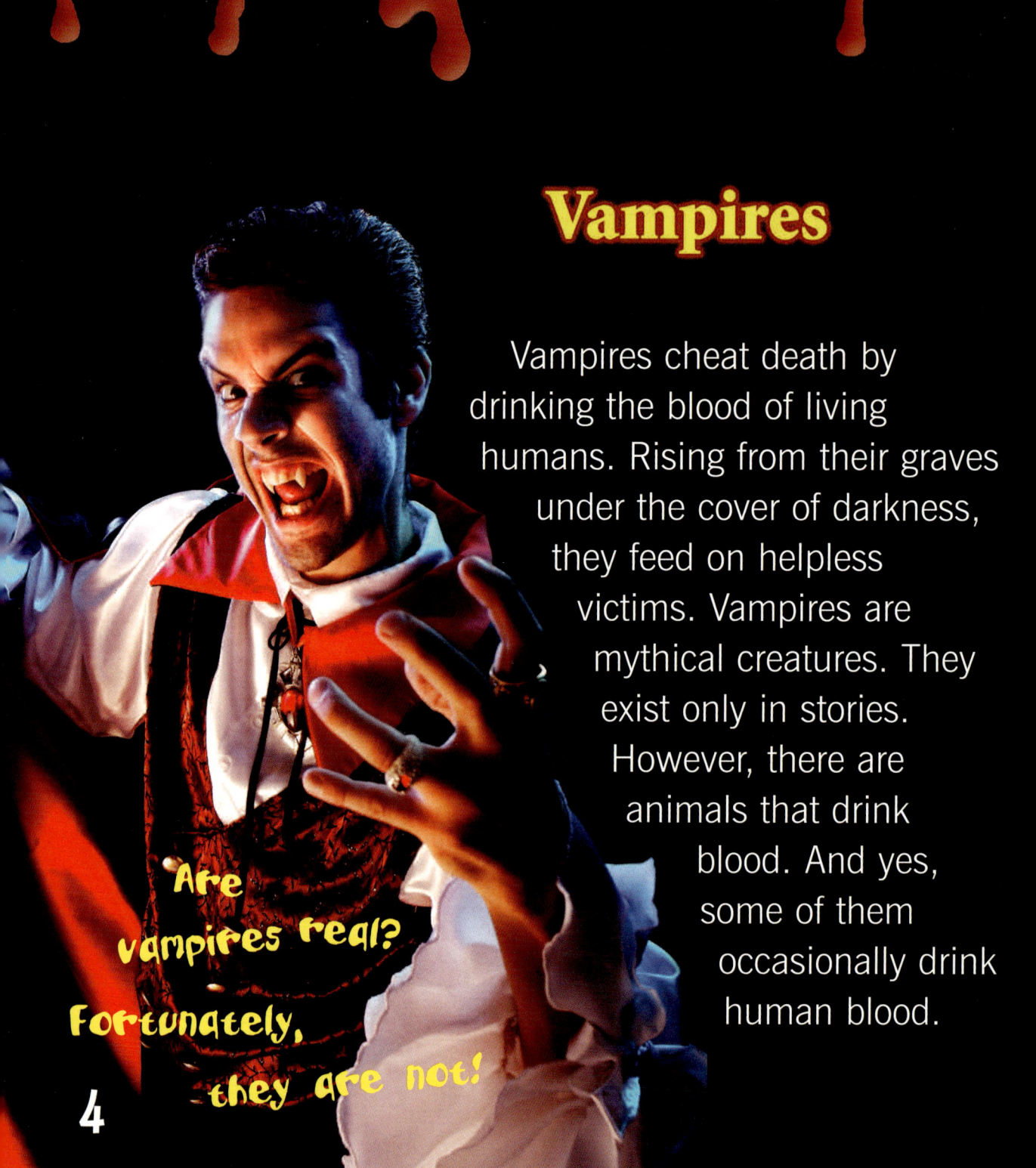

Vampires

Vampires cheat death by drinking the blood of living humans. Rising from their graves under the cover of darkness, they feed on helpless victims. Vampires are mythical creatures. They exist only in stories. However, there are animals that drink blood. And yes, some of them occasionally drink human blood.

Are vampires real? Fortunately, they are not!

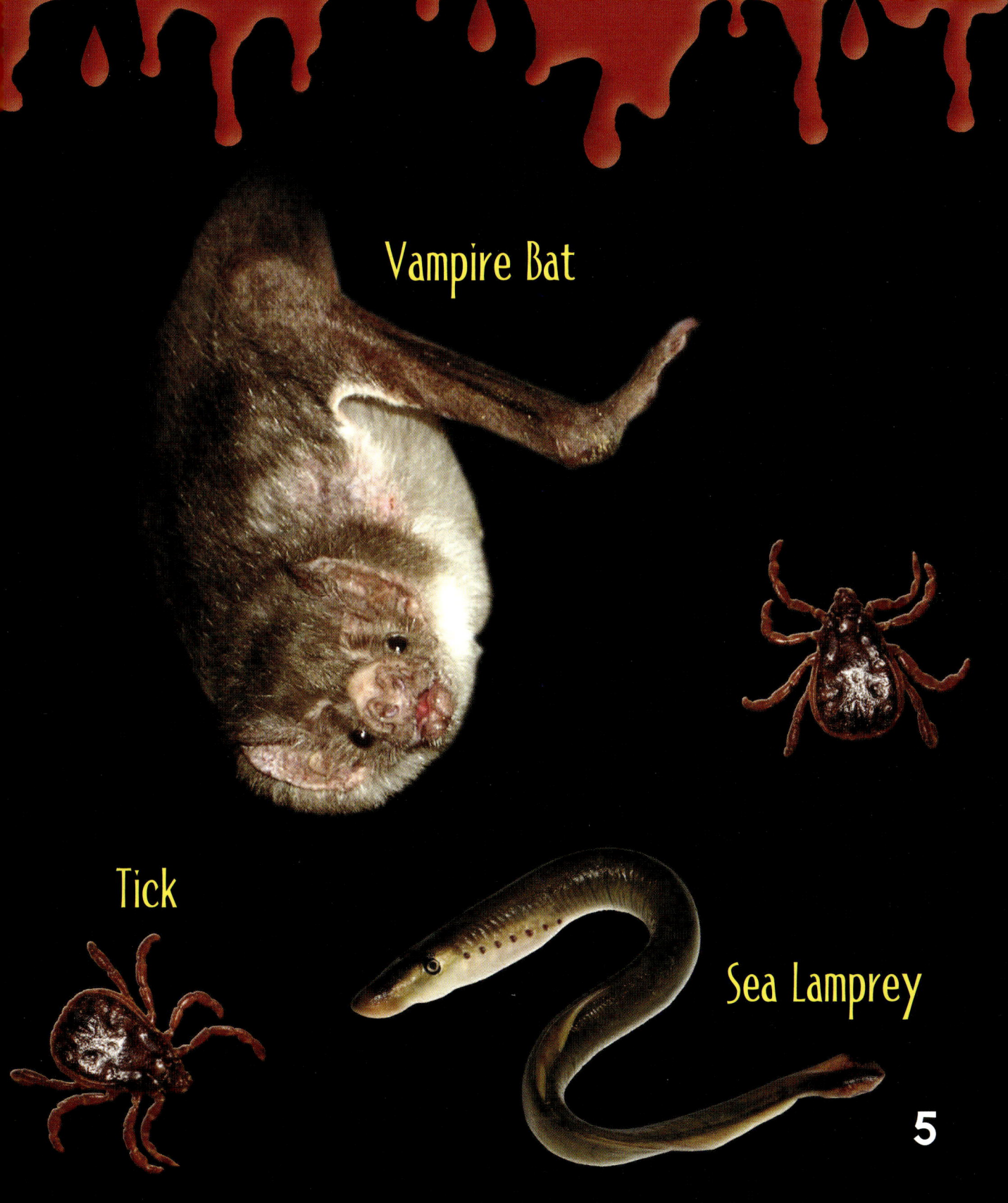

Bloodsucking Worms

Leeches may not seem as scary as vampires, but they are real bloodsuckers. Leeches look a bit like fat earthworms. Most live in water, but some **species** manage to survive on land in warm, humid environments.

A leech's body has 34 segments, each consisting of several rings.

A bloodsucking leech has a sucker on each end of its body. The suckers latch onto unsuspecting **prey**, sometimes human! Once firmly attached, tiny teeth painlessly slice into the victim's skin. The hungry leech draws blood through the tiny wound. A leech can drink many times its own weight in blood. When full, the leech relaxes its grip and drops away. Most victims never know the leech was there.

A satisfied leech may not eat again for months.

Long ago, doctors used leeches to drain blood from sick people. Many believed the practice, called bloodletting, had healing power. Thankfully, doctors no longer think bloodletting is useful. Leeches, however, still occasionally play a part in modern medicine. In some cases, doctors purposely attach leeches to a patient's wound. Leeches draw blood through the wound increasing circulation and improving the process of healing.

Not all leeches are bloodsuckers. Some are **predators**. Predatory leeches find tasty earthworms and snails and swallow them whole!

A leech attaches itself to its prey with strong suckers.

Vampire Bats

Everyone knows bats drink blood, right? Actually, only vampire bats drink blood. Most bat species are **insectivores**. They eat insects and are harmless to humans.

There are hundreds of species of bats, most pose no danger to people.

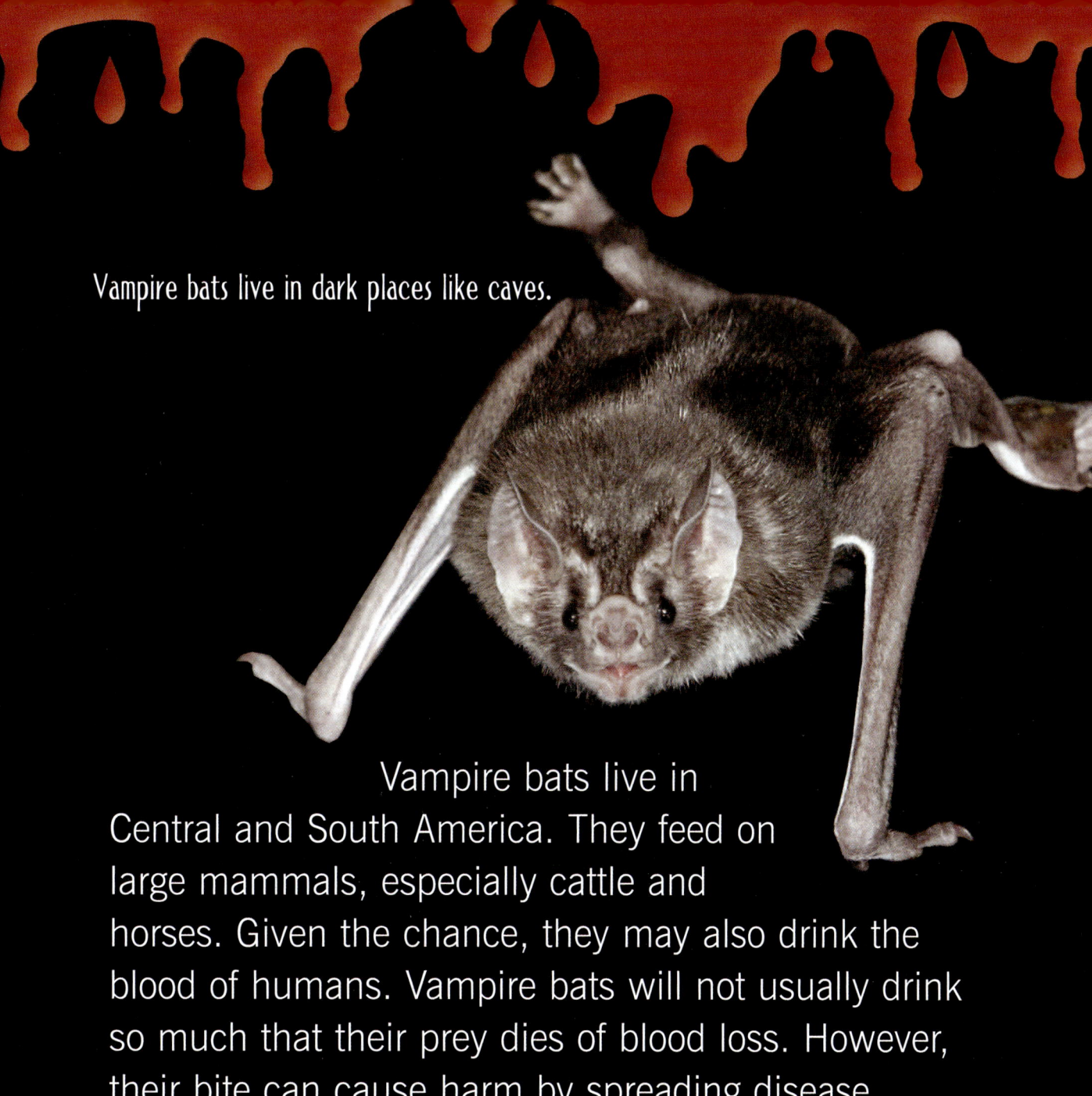

Vampire bats live in dark places like caves.

Vampire bats live in Central and South America. They feed on large mammals, especially cattle and horses. Given the chance, they may also drink the blood of humans. Vampire bats will not usually drink so much that their prey dies of blood loss. However, their bite can cause harm by spreading disease.

Like fictional vampires, vampire bats hunt only at night. Vampire bats fly, but can also walk, hop, and even run to pursue their prey.

Two vampire bats feed on a farmer's cow.

After finding its prey, preferably sleeping, a vampire bat quietly climbs on the animal. With razor sharp teeth, it cuts a small hole into the victim's skin. As the animal bleeds, the bat laps up the blood. Sometimes, when the bat finishes, another one takes its place and feeds on the same wound.

> Vampire bats have strong social bonds and live in colonies. A bat that has had success hunting will vomit blood and share it with other hungry vampire bats.

Calling Dr. Maggot

Flies lay their eggs in manure or in the decaying tissue of dead animals. When the eggs hatch, they are maggots. Young maggots live and feed on rotting flesh. Maggots are not pretty. They look like tiny little pieces of wiggling rice. That's enough to make your stomach turn, but it's not all bad. Doctors have found a way to make use of the maggot's disgusting habits. This is how it works: Doctors place **disinfected** maggots on a patient's wound. The maggots eat only the dead and diseased tissue. In the end, it is beneficial to both parties. The maggots get a free meal and the patient's wound heals faster. It sounds gross, but it works!

Maggots feed on the wound of an animal.

A fly prepares to lay its eggs.

Sucker for a Pretty Face

Sea Lampreys are jawless fish that look like eels. They are predatory fish that live in the Atlantic Ocean and its tributaries. They also recently invaded the Great Lakes, thanks to a man-made **canal** that gives them access from the ocean. Sea Lampreys are **parasitic**, meaning they live on and eat the flesh of other animals, mostly fish.

A sea lamprey can grow up to 35 inches (89cm) long.

Small teeth fill the mouths of two sea lampreys.

Sea Lampreys connect themselves to prey with powerful suction cup-like mouths. While attached, they feed on the meat and body fluids of their hosts. Tiny sharp teeth and rough tongues scrape skin and soft flesh from prey. Eventually, the victims either die or are released. Many large fish in the Great Lakes have round scars on their bodies from encounters with lampreys.

Two lampreys refuse to release their grip on a host fish.

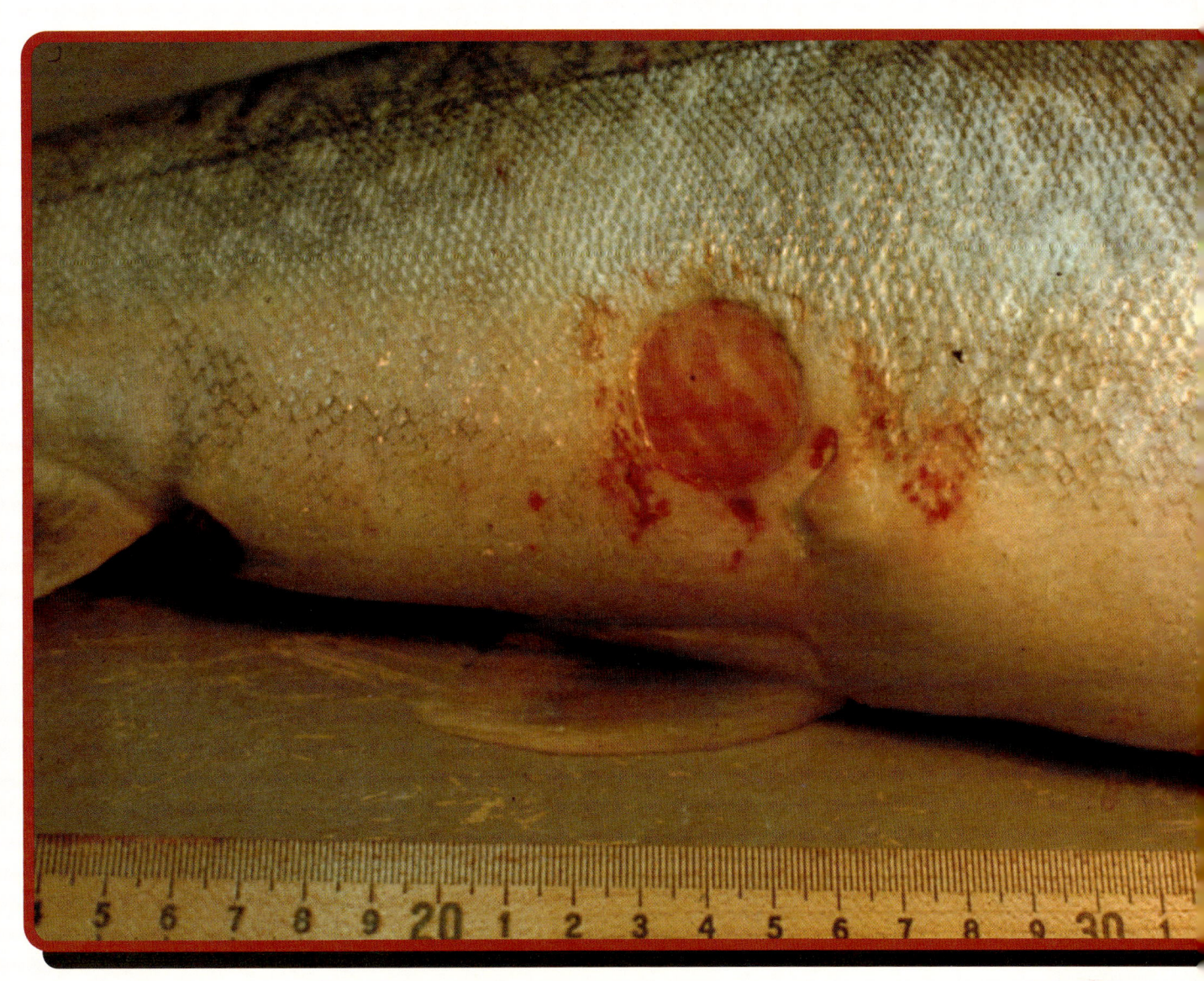

A sea lamprey's victim may die from blood loss.

Ticks

Ticks are tiny little pests that feed on the blood of mammals and birds. Ticks are **arachnids**, like spiders and mites. Ticks do not fly or jump onto their victims. Instead, they perch near the tops of grasses or shrubs. When a mammal or bird brushes past, they quickly drop or crawl onto their moving meal.

A lone star tick patiently waits for its next meal.

20

A tick fastens itself by digging into the host's skin with a barbed mouthpart. Once attached, they are difficult to remove. Their body swells as it sucks and gorges on the blood. A tick bite is not painful, but like other pests, ticks can spread disease to animals and people.

A magnified image shows a blood-filled tick.

Deadly Bloodsuckers

Maybe you've never seen a leech or a vampire bat, but how about a mosquito? Mosquitoes are small flying insects. They are common in warm **climates**. These uninvited guests ruin outdoor events everywhere they go. They buzz around ears and eyes and leave itchy bumps on skin when they bite. In some places mosquitoes are more than just annoying. They are deadly. Scientists say mosquitoes are responsible for the deaths of more people than any other creature!

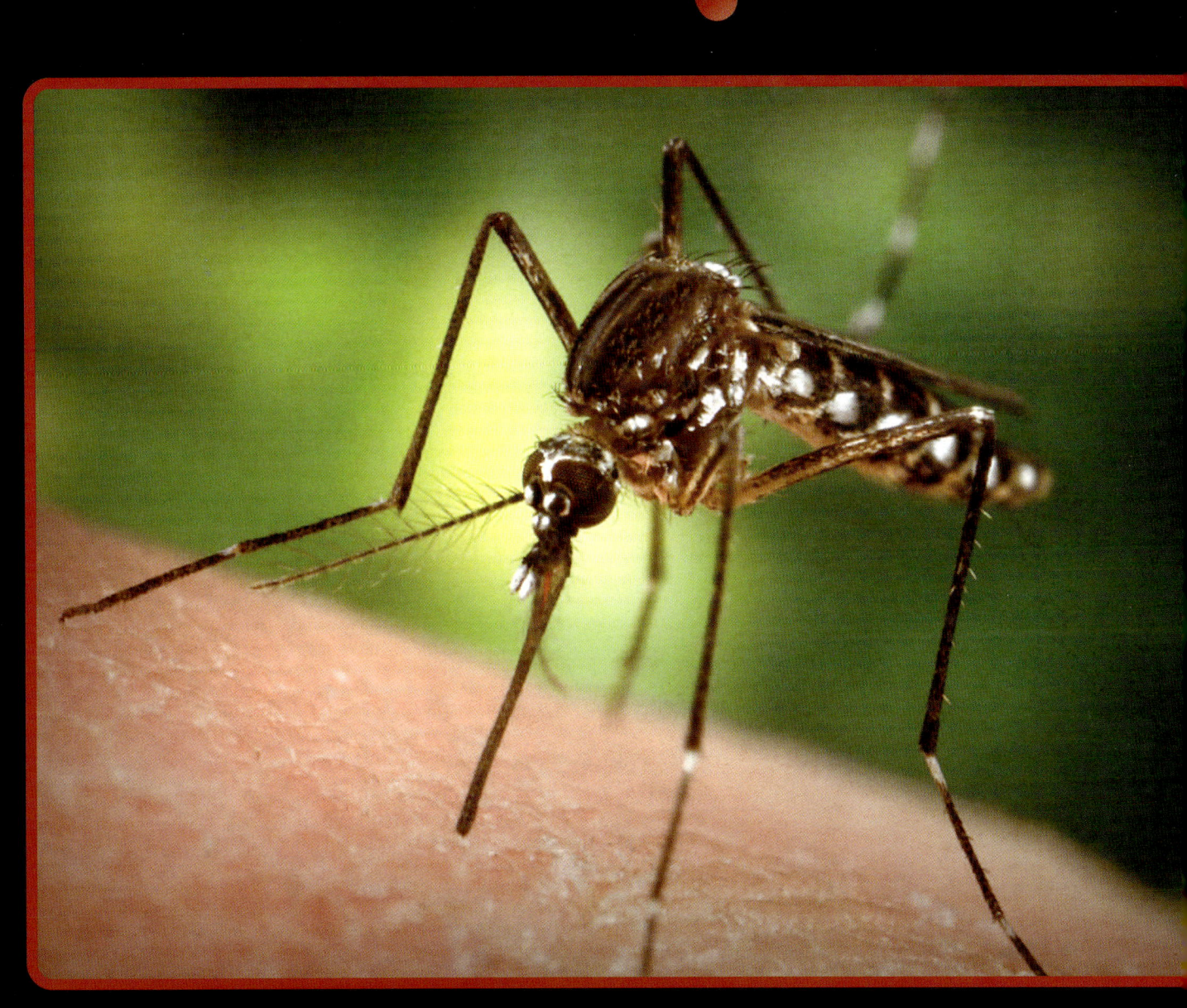

Female mosquitoes feed on the blood of other animals.

Mosquitoes kill people by spreading diseases like **malaria** and **dengue fever**. In parts of the world where medical care is hard to find, thousands die each year from infected mosquitoes. A mosquito feeds by piercing the skin with a **proboscis**, a hollow dagger like mouthpart. It uses the proboscis like a straw to suck blood. Mosquitoes spread disease when they transfer infected blood from one person to another.

A mosquito's abdomen is blood red and bloated after feeding.

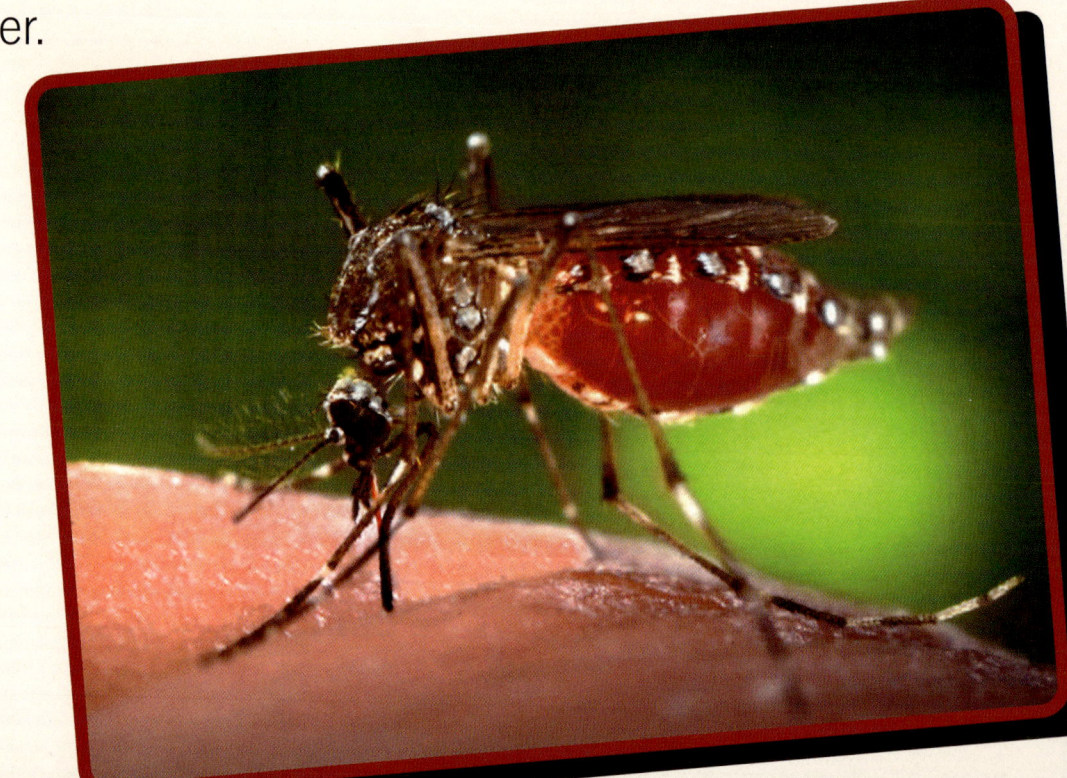

Chemical insect repellents keep mosquitoes from biting.

Only female mosquitoes suck blood. Males feed on flower nectar.

Don't let the Bedbugs Bite!

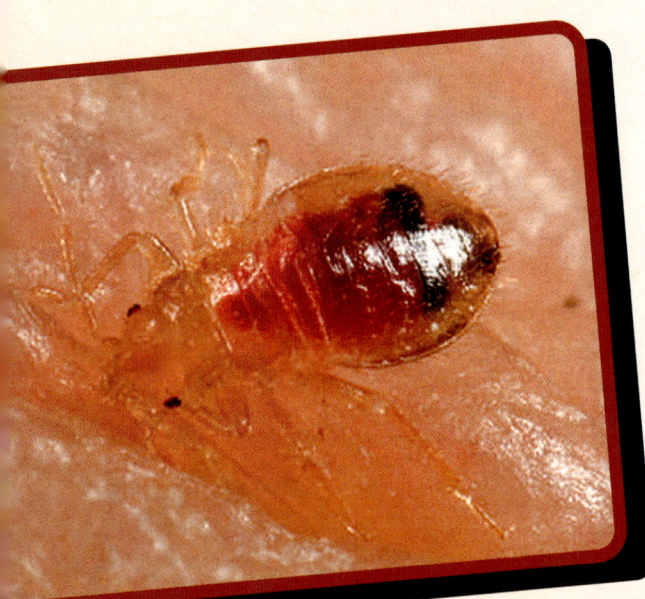

A newly hatched bedbug gets its first taste of blood.

Warning! This may not be the best thing to read before bed. Bedbugs feed on our blood while we sleep at night. Now for some good news, there really isn't much reason to worry.

Adult bedbugs turn reddish brown as they grow.

Bedbugs hide during the day and come out to feed at night. Their mouths have two sharp tubes that can pierce a victim's skin. One tube pumps saliva in to keep the victim's blood flowing. The other tube sucks out blood. Thankfully, bedbugs are tiny and cannot remove much blood. Their bites are not painful, but can leave itchy red welts on the skin.

An unexplained red welt might be a sign of a bedbug bite.

Killer Bugs

Assassin bugs have killer reputations for good reason. They are true predators. The name assassin even means killer. Assassin bugs pierce their prey with a long proboscis and inject poisonous saliva. The saliva liquifies the victims' insides, which they then suck out. Some species feed mostly on the body fluids of other bugs. Others prefer blood, and prey on larger animals. Assassin bugs do not usually feed on people, but they may occasionally bite them.

There are thousands of species of assassin bugs.

An assassin bug makes a meal of an ant.

Tools of Nature

The idea of an animal sucking your blood sounds like a nightmare. The truth, however, is not quite so frightening. Unlike vampires in horror stories, most bloodsuckers pose little threat to people. Mosquitoes are the biggest exception. They present a real danger to people in many parts of the world. Scientists are working to find solutions to mosquito problems.

Bloodsuckers are not evil creatures. They just use the weird and wonderful tools nature gave them to survive.

Glossary

arachnids (eh-RAK-neds): animals including spiders, scorpions, mites, and ticks

canal (kuh-NAL): a channel connecting bodies of water

climates (KLYE-mitz): usual weather conditions

dengue fever (DENG-gay FEE-vur): disease spread by mosquitoes

disinfected (diss-in-FEK-tud): cleaned

insectivores (in-SEKT uh vorz): animals that eat primarily insects

malaria (muh-LAIR-ee-uh): disease spread by mosquitoes

parasitic (PA ruh-sit ik): living and feeding on or inside another animal

predators (PRED-uh-turz): animals that hunt other animals for food

proboscis (pruh-BAHS-is): an animal's long snout or feeding tube

prey (PRAY): animals that are hunted by other animals for food

species (SPEE-seez): one certain kind of animal

Index

arachnids   20
assassin bugs   28, 29
bedbugs   26, 27
dengue fever   24
earthworms   6
eel   16
flies
leech(es)   6, 7, 8, 9, 22
maggot(s)   14, 15
malaria   24
mites   20
mosquito(s)   22, 23, 24, 25, 30
proboscis   24, 28
sea lamprey(s)   16, 17, 18, 19
spiders   20
tick(s)   5, 20, 21
vampire(s)   4, 6, 12, 30
vampire bat(s)   4, 5, 10, 11, 12, 13, 22

Further Reading

Gallagher, Belinda. *Animals: 1000 Facts*. Miles Kelly Publishing Ltd, 2007.

Mcghee, Karen. *Encyclopedia of Animals*. National Geographic Children's Books, 2006.

Somervill, Barbara A. *Vampire Bats: Hunting For Blood*. Powerkids Press, 2007.

Websites

http://kids.nationalgeographic.com/

http://kidsgowild.com/

http://pbskids.org/krattscreatures/login.shtml?

About the Authors

David and Patricia Armentrout specialize in nonfiction children's books. They enjoy exploring different topics and have written about many subjects, including sports, animals, history, and people. David and Patricia love to spend their free time outdoors with their two boys and dog Max.